Robotic Process Automation Tools, Process Automation and their benefits

Understanding RPA and Intelligent Automation

Srikanth Merianda

TABLE OF CONTENTS

Understanding RPA 3
    What Is the Definition of RPA? 7
    Various Types of RPA 8
    How Does RPA Work? 10
    Benefits of RPA: 12
    Better Business 13
    Better Productivity 15
    Better Economies 16
    RPA For the Back Office 18
    Why You Should Adopt RPA For Your Business? 27
    How the Workforce Will Be Redefined 30
    Addressing Common Concerns 33
    Current RPA Vendors 35
    The Future of RPA 41
References 43
More Books from The Same Author 44
About Author 45

Understanding RPA

All companies, big or small, look out for profitable ways to provide faster and better service to their customers. In an increasingly competitive world, there is a real need to do more with the resources available. Hiring manual labor to complete repetitive work, like data entry, is a tedious and expensive solution. In current times, manual labor is quite uneconomical, slower and often erroneous.

A few decades back, outsourcing back office and operational work to developing countries became the new norm and trend. That is the reason why cheap labor from developing countries became desirable and quite sought after. The BPO industry was booming in many parts of the world. However, with rising labor costs and constricting regulations all over the world, outsourcing to quality cheap labor is no longer viable for companies. That is when AI became a glimmer of hope, leading to a different path in a profitable way work gets done.

With the advent of advanced technology, more companies are shifting towards robotic automation. They do so because the benefits of implementing RPA considerably outweighs its drawbacks. RPA provides a comprehensive virtual workforce that is both efficient and reliable. Companies looking to quickly scale their operations, but haven't yet turned to automation, should seriously consider opting for RPAs.

Process automation has been around for longer than we care to admit. It is the robotic part that is advanced and hi-tech. Inventions like the steam engine, water wheel and printing press automated work that was previously done by humans. Just like the way those inventions revolutionized the way the world functioned, RPA has just as much power and potential to do so. And the best part is that we are at the cusp of this very real industrial revolution.

Often touted as the fourth industrial revolution, AI, and by extension RPA, are changing the way companies operate across the world in a drastic and unique manner. However, the evolution that succeeds every industrial revolution demonstrates this. The first pioneering industrial revolution automated processes by utilizing the power of water and steam to enable the mechanization of production processes. The second industrial revolution utilized electric power or electricity to enable the mass production of various goods. The third industrial revolution introduced the internet into the business world. The internet was also called the digital revolution and brought about the Information Age. The fourth and current revolution we are undergoing aims to change the way we work with artificial intelligence (AI) and the Internet of Things (IoT). The robotic process automation (RPA) uses artificial intelligence (AI) to integrate and automate the work processes, thereby redefining the way we get things done – yet again. Hence, it is being called an industrial revolution.

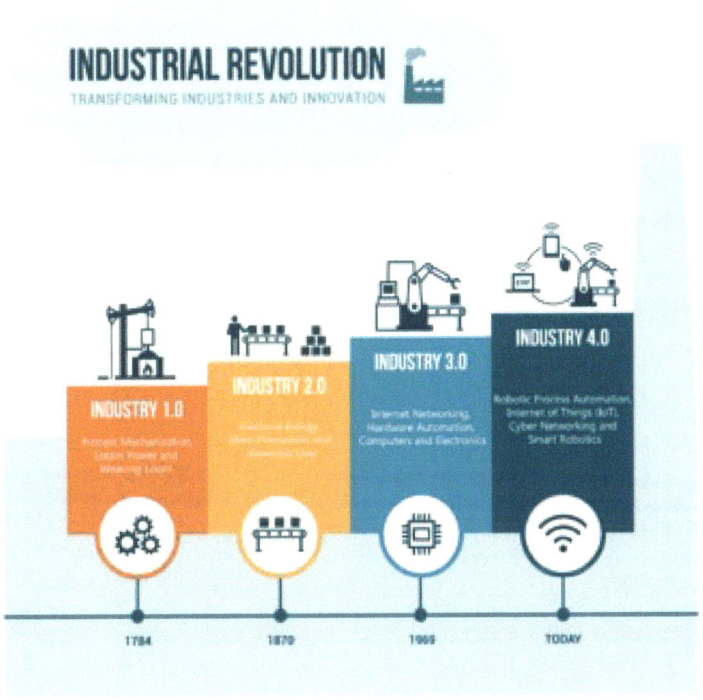

While it seems that the fourth industrial revolution is an extension of the third, it is really not. Sure, AI is a reality right now only because of the internet. But that doesn't necessarily mean that AI is not an industrial revolution in its own right. It is actually quite the opposite. Much like the industrial revolutions that came before, AI is changing and revolutionizing the way we do things. From this point, there is really no going back. It is only going to get bigger and better, much like the same way the previous industrial revolutions affected the way we worked, and then soon afterwards, affected the way we lived.

A remarkable manner in which the fourth industrial revolution is unique is that the growth in terms of advancement has been rapidly fast, exponential even. In the case of the industrial revolutions that came before, where it took some time before global acceptance, their growth and complete adoption went on in a slower pace. However, the current industrial revolution has already reached almost every part of the world. Keep in mind that the internet was publicly available for mass public use during the late 60s. Before the internet, the second industrial revolution took place during the 1870s. Before that, the first industrial revolution took place during the 1780s. There is almost a century's worth of gap between the first, second and third industrial revolution. However, not even two decades after the internet was made public, the fourth industrial revolution is already underway. The strides AI has leapt is astonishingly swift, ensuring its survival and adoption in a way that could be aptly termed as 'sooner rather than later'.

What Is the Definition of RPA?

RPA or Robotic Process Automation is the path to a better future. Robotic process automation are algorithms or configurations that automate high-volume manual and repeatable tasks. They are designed to solve specific problems, so you will need different RPA algorithms to solve different problems.

RPA solutions can be easily and quickly designed, tested and implemented. An experienced software developer can build efficient RPA systems so they require a relatively low amount of investment or expenditure.

RPAs, often called bots, imitate a human employee. They work by complementing a company's infrastructure, logging into the company's software, doing the work required, and logging out. This means that businesses can implement RPA to do the required work without having to change their infrastructure.

RPAs can take care of laborious jobs, like data entry, that form the backbone of any business. These algorithms are reliable and efficient, which means that the work will be done accurately. RPA produces faster and cheaper results with better accuracy. This also means that businesses can get work done on tighter timelines and valuable human resources can be used elsewhere, in a productive manner.

Various Types of RPA

Robotic process automation is a broad term used for the practice of involving software or machines into the workforce. It comprises of algorithms that solve a specific set of problems. So, for different kinds of problems, you need different kinds of RPAs. The sophistication of RPA algorithms is determined by looking at the kind of work that it needs to do. Basically, there are two kinds of RPA: hardware and software.

Hardware RPA are the kind of robots you see in factories. These robots make the process of building objects faster, safer, easier and more accurate. The consistency of hardware robots is quite desirable in many industries ranging from food to automobile.

The other kind of RPA is software. Software RPA is a set of coded commands used to run programs that is designed to help in completing tasks. It runs on a host device like a server, instead of running on a standalone machine. The different types of software RPA can be best be broken down into three kinds:

1. Basic Process Automation:
   RPA was first built for transactional, rules-based tasks where structured data is involved. It is designed to follow clear, pre-defined rules and parameters to execute

such tasks. Software robots that handle structured data like data in spreadsheets, CSV and XML files all come under basic process automation. This is especially useful to companies who want to automate jobs like data entry.

2. Enhanced Process Automation:
   Enhanced process automation enables software robots in the use of structured and unstructured data to support self-learning. The software robot can understand unstructured data, human communication like voice or email, and draw conclusions from cross-checking the data. This kind of RPA can be useful to companies who want to automate jobs and introduce virtual assistants and office secretaries.

3. Cognitive Robotics:
   This kind of RPA provides decisional support due to the advanced algorithms automating processes that are more cognitive in nature. Cognitive robotics involves programming a robot with intelligent behavior by providing it with a processing architecture that will allow the robot to learn and reason about how to behave and respond to the complex goals or tasks. Cognitive robotics empowers robots to learn and complete complex work without much human interference. An example of cognitive

robots are chatbots that intercept customer queries.

These are the three broad categories of the types of RPA that exists. Often, to tap into RPA's potential, companies will need to deploy a combination of these three kinds of RPA. When doing that, businesses have to be careful as to tailor the deployment in a way that will benefit their process. When undertaking a major shift like automation, it is best to consult an IT specialist who has experience in guiding and handling the change to automation.

How Does RPA Work?

The best way to explain how RPA works is by citing an example. In this example, I will first begin by detailing the task that needs to be completed. Then, I will explain what a human employee would do to complete the task. And the, I will compare that with how RPA does the same task.

The Task:
The client sends a request for an invoice to be raised. The client sends this request via email. This request reaches the company ERP (Enterprise Resource Planning) software. Now, an accurate invoice needs to be sent to the client.

The Human Execution:
A human employee logs into the correct division in the company's ERP. They notice an invoice request is received via email. They open the email and the attached excel document. The employee now verifies the information in the excel document. If there is some inaccurate or missing information, the employee sends an email requesting the client to adjust/ add the missing or inaccurate invoice data.

Once the client sends through all relevant information, the employee copies the information into the ERP software. After that, they raise an invoice with all the correct data typed into it. Then, after saving a copy in the ERP software, sends an email to the client informing them that the invoice has been created. After that, the employee closes all the used applications correctly.

The Automated Execution:
The RPA "bot" opens the correct division in the company's ERP and logs in. The bot receives an invoice request from the client via email. The software robot opens the email and the attached excel document and scans it meticulously. The algorithm then conducts different checks to verify whether the data in the excel document is accurate. If everything checks out, the bot runs the transaction code to create the invoice in the ERP software.

After that, the invoice data from the excel document is

copied into the relevant fields in the ERP software. When there is essential information missing in the invoice request, the bot immediately recognizes it. If this is the case, the bot sends an email to the client requesting them to adjust/ add the missing information into the specific invoice data.

When all the data is entered into the ERP software, the bot creates the invoice by running the correct transactions. After that, it correctly saves the invoice for sending. At the end of the process, the bot sends an email to the client with confirmation that the invoice has been created. At the end, it correctly closes all the used applications.

Benefits of RPA:

There are many tangible benefits to implementing RPA. Automating work is a great way to make those processes better. Robotic process automation helps reduce human errors dramatically. This means businesses can increase their work hours and use their human resources to focus on tricky problems that actually require human intervention.

While humans require rest, tire easily and need vacations, robots can work continuously 24/7. Human error is quite common, and after all, to err is human! But robots don't make mistakes, making their work highly accurate. When a company wants to scale its business, they need to hire more workers. With RPA, they just need to install more systems. RPA is also very scalable and can be scaled without much effort or expenditure.

Especially for those in a highly regulated industry like finance, there are a lot of compliance costs. Compliance costs are the expenditure of time or money spent to conform with government requirements, regulations and legislation. This is because with human workers, the job need to be executed and then saved, stored or recorded. This consumes time, effort and money. RPA automatically stores and records everything it executes. There is no need of additional effort or expenditure. This translates to easier documentation for legal and tax purposes.

Better Business

According to a study by Deloitte, after implementing RPA, there is a 40% decrease in the average process execution time and 45% increase in the ability of employees to focus on customers. Process execution time is the time taken for a business to complete any particular work. The increased productivity would give the company opportunities to expand their clientele. And with more number of clients, companies can use their human resources to tend to customers in a better and helpful manner.

RPA dramatically reduces costs also because companies can afford to downsize their workforce. Once RPA has been implemented, those processes gets cheaper to execute. Companies get to do a large number of processes cheaper and faster, without having to pay a huge group of human workers.

Another bonus about the way RPA works is that it can handle a sudden increase in the amount of work. Human employees can only work for so long before they get tired (or till it goes against the labor laws!). More importantly, if in case there is a reduction of work, companies won't go at a loss for that period. If human employees were in that situation, even if there was not much work, they will have to be paid. During the occasional lull, there doesn't have to be any kind of expenditure on RPAs.

In terms of security that leads to better business, a software robot can be used to execute a process as instructed, without incidents of inappropriate data collection, fraudulent intervention, data leak or deviation from the stipulated process. Insider trading with the most sensitive data is also not possible with a highly secure RPA.

Better Productivity

RPAs execute the required work more efficiently than the average human employee, without breaks or wastage of time. RPAs don't get bored or tired. Bots don't take personal time or holidays off. This means the company gets more work done during any given time. With more work being completed, businesses make more money. These remarkable advantages would benefit any business.

Also, when human employees fall sick, there is no exact time frame on when they can get back on their feet. This hampers the company's productivity and efficiency. However, with a robot, they don't fall sick or hamper productivity. Even if they become unfunctional, you can hire an IT consultant to fix the problem within a set time frame and your company's operations are back on track.

Every job has requirements that are unique to the company. No two jobs can truly be alike So, a human employee needs to be trained to work in a way that fully benefits the company. An RPA doesn't have to be "trained". All that is needed are configurations and calibrations to tweak the RPA functioning, and the RPA works in the way where your business benefits. There are no huge employee training costs when using RPA. Alternatively, the company does not suffer if the employee is ill-trained.

Better Economies

Sure, the advantages mentioned above does correlate to better economies. But in this sub-section, we will solely focus on the core financial benefits a company will reap if they opt for RPA.

The Deloitte study also states that after implementing RPA, on average, businesses achieve a positive payback within three months. It mentions that the initial investment spent on implementing RPA can be gained back within three months, on average. Automated processes can do work for much longer periods of time, when compared to human employees. Also, the cost to run RPAs is much lesser than the monthly salaries given to human employees for doing the same work.

Another benefit is that RPA is modular. Imagine you want to restructure your business. You want to shift one department to one end of the building and another department to another end. In the usual scenario, you will have to re-build the two sections of the building to provide appropriate facilities and amenities for human workers to work there. With RPA, however, as it is server-oriented, you just need to shift the systems. That's it – without much effort, you get to restructure your building.

There is also the topic of safety standards. While no

business can solely run with robots, the company does spend a lot more when they have to comply with the statutory safety standards. Instead, after the introduction of RPA, the expenditure on enacting safety standards is reduced and the company can use that money in a better area like employee satisfaction.

Real estate costs are quite high. Unless your company works in a far-flung area, real estate costs eat a large chunk of capital. Employees need a good amount of real estate to work in. RPAs, however, don't need as much. By automating clerical processes like data entry, businesses can save a lot of money by reducing their need for real estate.

RPA For the Back Office

In this section, I will detail on why RPA should be implemented for back office operations – no matter what kind of company and what type of industry they are catering to.

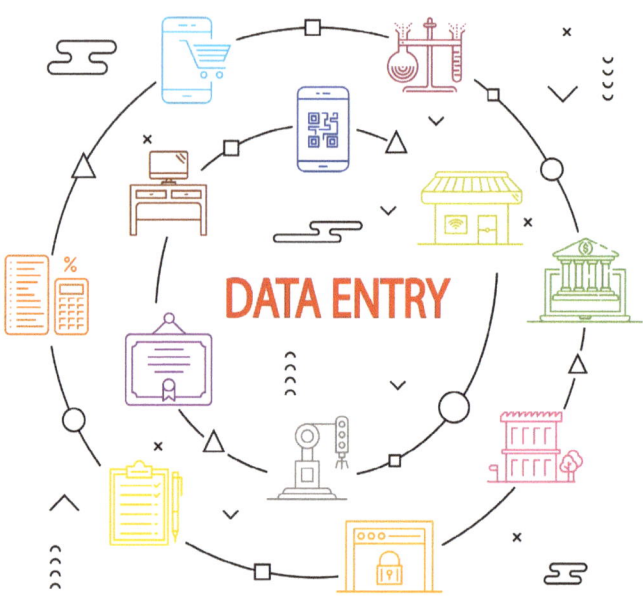

The back office is an important part of any company that consists of administration, maintenance and support personnel. The backbone of any business, this aspect of the company does not face clients. Some of the operations and functions that are categorized as back office operations are payment settlements, due and credit clearances, record maintenance, legal compliance, accounting procedures and IT services.

However, it is the back-office operations that directly influences client experience. For example, a customer service representative can only offer excellent service everyday if the details of the customer are typed into the system accurately. It is up to the back-office employees to do that. If the data is inaccurate, customer experience will not be excellent. It doesn't matter to the customer if human error was the cause. Finally, the customer's experience with the company sours and this can have serious repercussions. However, more often than not, human errors can mar the road to excellent customer experience. If it happens on a large scale, this can lead to a bad reputation and even the downfall of a company.

Great customer experience should be at the forefront of every company's priority, and rightly so. Only if the customer experience is positive will customer loyalty stay strong. This leads to more business and a better brand name for the company. No matter what industry the company caters to, back office operations is one of the major weak links. This is more evident as the company grows bigger. As the data that needs to be processed increases, the company has to hire more human employees to finish the tasks. The nature of back office operations is that it is repetitive and strenuous at the same time. Over time, human employees can tire from the tedious and mundane type of work. This leads to human errors and inefficiency. Especially in the case of invoice processing and account payables, one small error can hugely impact and affect your business.

By implementing automation to back office operations, software robots can do the repetitious and tedious work, while human employees can focus on tasks that lets them add value to initiatives that are oriented towards customers. RPA ensures that there are no errors, with the right calibration. Software robots don't get fatigued and work much faster than a human could. In areas like data entry and invoice processing, speed and accuracy are extremely important. The best part is that robotic process automation can deliver exemplarily without being too expensive.

The back office is present not just in companies of different industries, but also inside different departments in any company. As a company grows, to handle the increasing amount of input of various kinds of data, the back office of the company grows as well. Here are a few examples where RPA is very useful in different departments and branches in any company's back office.

1. HR Department:
   Data entry in the HR department is usually a lengthy and tedious process. Major functions in the HR department includes keeping a record of the incoming job applications, employee

complaints, tracking employee performance, grief handling, exit interviews, recording interview data and filing/ updating details of employees into company's MIS. These are just some of the areas in the HR department that would require data entry.

Especially in a big company where there are many employees, a human employee would take hours to enter so much data into the company's systems, even more so if the company size is larger. On the other hand, RPA can do the same work much faster. Automated systems can also remind when a particular employee has to be paid – all without any hassles.

2. Supply Chain Management:
   Many components in supply chain management requires data entry. Whether it is integrating functions, coordinating processes, checking inventory or managing resources, it all relies on the data being entered into the system. Also, when successfully handled, many elements that make up the supply chain like taking care of inventory deficiency, logging in inventory data, noting inventory deficiency differential, tracking delivery time and tracking goods all ensure goods move smoothly.

However, since there are many constituents that make up the supply chain, there are also many areas where things could go wrong. Nowadays, most of the manual process in supply chain management are handled by RPA systems. This leads to less erroneous data and more accurate, and therefore smooth functioning, of supply chain systems.

3. Finance & Accounting:
   The accounting department is key to any company. Entry of data regarding invoices, credit loans, unpaid balances, tax information, compliance data, etc. are all crucial to the proper functioning of any business. Even for areas of customs, taxation, costing, any error in the data could lead to massive legal and financial troubles for the company. With RPA, entering data into the system would make the process smooth, fast and, most importantly, accurate. The percentage of error drastically falls when the same work is done by an automated software robot.

4. Pricing, Sales & Marketing:
   Quite often, marketing teams embark on their campaigns based on market research, data and analytics. Even in the case of sales teams, the

way they work depends on data. Collecting and analyzing this data takes a lot of time and effort, which leads to delayed results, delayed revenue, and potentially, a lot of missed opportunities. If we delegate the research, collation and analyzation to RPA, we could get useful information faster which will help in closing deals and bringing revenue faster. This is the sales aspect.

For the marketing aspect, getting relevant data like consumer trends, preferences and patterns faster would yield in churning out marketing campaigns that are quick, robust and effective. We could also make software bots troll and find business leads and insights about consumers and the competition. The best part is that RPA can function 24/7, so businesses can find and crunch data, obtain genuine leads and potential clients continuously. All this information can help the sales & marketing teams become much more productive, efficient and successful in their efforts to secure more business for the company.

As for pricing, it is closely intertwined with any company's marketing and sales. Pricing enables the product to be desired and bought, fulfilling both aspects of marketing and sales. Only can a well-priced product be well received by the market. In the same fashion, only can a well-priced product be successfully bought by the market. To price a product in a way that ensures sustainable profit and sound reasonable to customers, pricing teams need a lot of data. Everything from consumer patterns, profit and loss, competition insights and market demand influences the pricing that will make your product advantageous. So, getting the right data swiftly is very important. RPA gets all of this done in a very fast, efficient and meticulous manner.

5. Customer Service:
   To provide excellent customer service, accurate data entry is paramount. Every customer would like speedy service, as customers consult a company's customer service only when they are distressed. When a customer asks for help or queries, it will cause irreparable damage if the customer's data is precise. Only if the data entered into the system is error-free, will the customer get the best service. Speedy and helpful service boosts customer loyalty. When customers don't get constructive service or solutions, no amount of training given to customer service employees will repair the

damage that has been done.

6. Maintenance and Repair:
   Maintenance and repair departments inspect and take care of the quality of machines, mechanical equipment, buildings and factories. There is a very big need to keep the back-end logistics and statistics of company assets error-free to avoid large scale corporate mishaps. Especially in big companies where there are many assets to take care of, the background checks done by a big team of maintenance engineers will result in a large amount of data that needs to be entered into the system. RPA does the same amount of work in a fraction of the time. Be it specifications, service history or logistics, RPA can quickly update the maintenance logs with information that is reliable, precise and secure.

7. Manufacturing:
   In a company owning factories, logistics and data entry acts as the manufacturing unit/s backbone. While RPA is used as hardware robots to build and create goods, in this section, we are talking about RPA to aid data entry. If the data is erroneous or misleading in any way, the manufacturing unit/s could fall under legal,

financial, manufacturing troubles, to name a few. An easy way to avoid such issues is to implement RPA to deal with the logistics and data. Many manufacturing units have implemented it and the results speak for themselves. Operational accuracy drastically increases while reducing flaws and shortcomings. While machines building a product does have its limitations, the incapacities of hardware robots can be truncated by software robots or RPA. There are many kinds of work that are still completed by humans like quality checks, updating, scheduling and reporting. These kinds of tasks can be outsourced to RPA systems that can smoothly and efficiently handle such processes.

These are just some of the major departments in a company where the back office can be automated. With a more robust, streamlined and brisk support, companies can expedite processes to make their functions and services better and faster.

Why You Should Adopt RPA For Your Business?

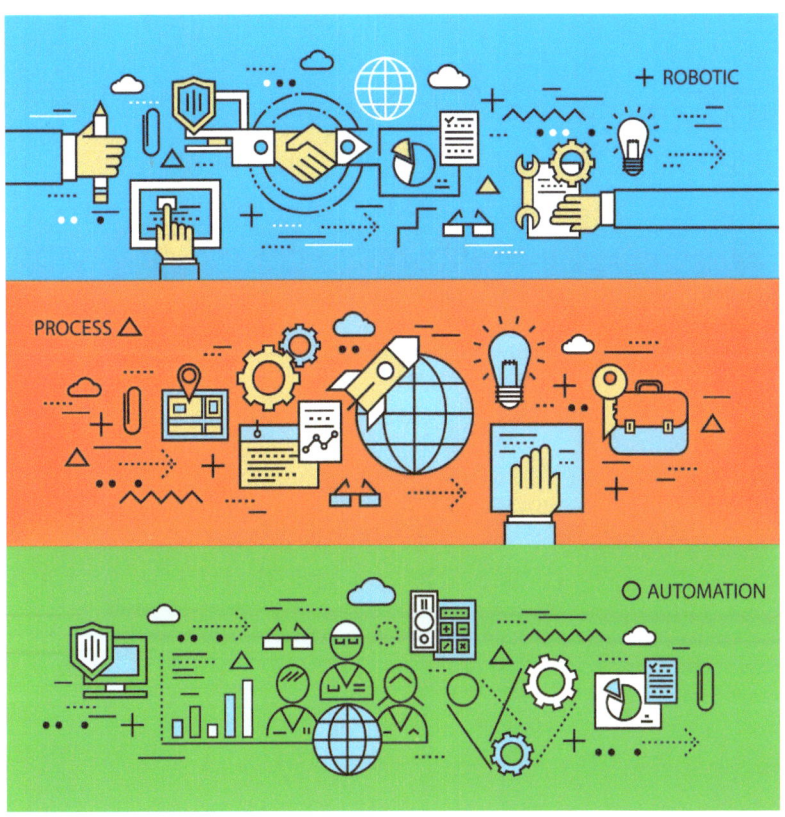

Today, every business strives and vies to sustain lean operations and delivering exceptional client experience while running at the lowest possible cost. A viable tried-and-tested method of achieving this is by successfully adopting robotic process automation into company operations. RPA ensures operational efficiency like no other solution. As operational efficiency is a priority for every company, and RPA solves this problem effectively, it is not hard to see that RPA is here to stay.

Your business can be positively impacted with the successful implementation of RPA. Only if the whole process, from identification to implementation, is customized to your company requirements and goals will the positive impact become a powerful one. It is strongly recommended that you consult an IT specialist to help guide you and your company to the path of success.

To adopt RPA effectively and productively, one must have a good understanding of specific business functions and verticals whose processes could radicalize with the implementation of RPA. To truly unravel the potential and capabilities of RPA application, one must stay true to the needs of the company looking to implement RPA. After all, RPA can transform the functioning of business processes, businesses should just embrace the change.

It can be hard to determine the areas where you need to implement RPA for better productivity, business and economics. It is important to identify key areas and business processes where RPA can make the biggest differences. Here are some features and pointers about RPA to keep in mind when looking to implement RPA into your business.

1. Higher Chance of Human Error:
   There are many areas in a company where tasks are monotonous, tedious and time taking. One of the areas that match this is data entry. When a human worker sits for hours doing the same laborious work for too long, it translates to poor quality and error-prone data that has been keyed into the company's systems. In such processes, RPA can finish the same work quicker and do it more accurately.

2. Clear Rules:
   RPA functions smoothly when there are clear instructions or rules on the steps that needs to be completed to finish the tasks. Since RPA is an algorithm, the rules have to be well-defined with a step-by-step process, with exceptions or anomalies that can be dealt by human employees.

3. Fewer Exceptions:
   An extension of the previous and the next point, RPA can realize its full potential in processes where the exception is not the norm. In many areas, human intervention is essential. In such areas, where RPA can be accommodated, RPA cannot be implemented properly or fully. So, fewer exceptions or anomalies that require

minimal human involvement are the areas where RPA can truly take charge of the work.

4. **Minimal Human Involvement:**
   The whole point of introducing RPA is to let the software automation take care of the work. So, RPA works best for areas where there is minimal human involvement, once the automation has commenced. However, there are multiple scenarios where RPA and humans can work together to reach the same goals.

These are just some of the basic attributes of the sectors and sections of any company where RPA can alter the way tasks are completed. The shift to RPA establishes streamlined processes which leads to faster and better business.

How the Workforce Will Be Redefined

Imagine a data entry employee. Their job is to enter data of customers into a log sheet, over and over again. In time, their efficiency and accuracy will reduce, because they do the same job repetitively. In comparison, a well-designed RPA will do the same work as the data entry employee, but this time around, it will be faster, more accurate and more efficient.

This is not to say that RPA will replace the human workforce; however, it will somewhat reduce it, streamlining the process and making it more efficient for the business.

RPA has already started redefining the workforce. Many companies that have adopted RPA into their systems are growing faster than ever. Companies that haven't yet espoused RPA are falling back. This is the reality. With the push to go digital, RPA technology can easily fit into your company – without disturbing the company's infrastructure.

With the introduction of RPA into the workforce, there are many changes the typical workforce will undergo. While not necessarily negative, there are many marked changes that will take place in the company hierarchy.

1. RPA will yield to different kinds of jobs:
   With the entry of robotic automation into the workplace, it will increasingly create more jobs outside the traditional boundaries. As far as traditional jobs are concerned, the job description will get more fluid. The standard, routine work will be taken care of RPA while the non-routine work will be taken care of humans. This restructuring of work will lead to new and atypical kinds of jobs.

2. Traditional jobs will be deconstructed:
   As RPA gets infused and blended with the traditional work that takes place, the job that was once completed only by human employees with be deconstructed to integrate RPA into it. For example, the job of a supply chain manager was to keep a check on logistics and purchase necessary inventory. Now, the logistics and tracking part is taken care of by software robots. The human employee takes care of purchasing and making suggestions that improve productivity, quality, and efficiency of operations.

3. Re-envisage the definition of an organization:
   Typically, organizations consist of bosses and employees. With RPA entering the realm, the definition of what an organization consists of is radically altered. With a variety of RPA that will join the workplace, a business will soon comprise of bosses, human employees, software RPA employees and hardware RPA employees.

RPA will remarkably disrupt and radically empower the global workforce. Company leaders should embark and draw up a nuanced automation strategy that perceive its benefits, evades needless costs, and rests on a foundation of a more nuanced understanding on the topic. While it won't transpire all at once or in every single job, it will surely take place leading to a massive paradigm shift in the way we all work.

Addressing Common Concerns

There are some concerns regarding RPA replacing the human workforce. Robotic automation does, to some extent, reduce the number of jobs for human employees. But, RPA can never replace humans. RPA is just an algorithm that does what it is programmed to do. They can't do anything more. For any situation that require nuanced handling, human intervention is essential. RPA can never be a substitute to humans. In a study conducted by Harvard, it states that, rather than an all-human or an all-bot approach, the human-automation combination provides optimum efficiency. This is because RPA can do clerical work faster and more accurately than humans, but the human touch is vital to ensure the that the work is done correctly.

Implementing RPA into companies is not just to reduce the costs of paying employees. Yes, there is some reduction in the number of salaries that needs to be paid, but there is so much more to it than that. By adding RPA into the workforce, there will newer kinds of jobs and the quality of work done will of higher quality. And with more money available to employers, they can afford to focus and increase employee benefits, which will directly lead to the increase in employee satisfaction. Businesses introducing RPA do so to make laborious and repetitive jobs, like data entry for example, more robust.

Rather than reducing employee satisfaction, the Harvard study states, employees embrace the change to RPA, and viewed the robots as "teammates". Also, to be noted from the study is that many companies adopting RPA have promised their employees that the change to automation will not result in layoffs. Instead, the companies decided that workers will be redeployed to do more interesting work. The same study acknowledged that, instead of replacing humans with robots, companies used the RPA technology in a way that achieves more work and greater productivity while retaining the same number of human employees.

For example, in the job of a call center representative, RPA can take over some parts of the job like requesting customer identification information and tracking the status of a delivery. RPA can hardly take over the entire job; customer interaction has to be done by a human employee. However, this will make the process robust for the company, faster for the customer and easier for the employee – resulting in a better business, happier employee and stronger customer loyalty.

While this particular claim is a little whimsical, robotic process automation cannot replicate human cognitive functions. RPA follows a set of rules and cannot do anything beyond what the algorithm instructs it to do. So, robotic automation can't use its own "brain" and definitely won't try to take over anything!

Current RPA Vendors

When you want to integrate RPA into your business, you should hire a credible IT consultant to properly advise and guide you through the entire process from choosing the correct vendor to implementing it in your infrastructure.

There is no shortage of RPA vendors to choose from. Each product claims to be the best for your business. Some of the popular RPA vendors are Automation Anywhere, Blue Prism, KoFax, Kryon Systems, Workfusion, Pega Robotics and UiPath. Each product has different features, and claims to be perfect for all types, kinds and sizes of businesses. Choosing an appropriate vendor for your company is a very important decision. There are a few key points to think about when choosing an RPA vendor.

1. Accessibility:
   One of the most important things is the software's ease of use. Only if the product can be used easily, can it be used to internally scale operations effectively. The product should be easy to handle, making it easy for your business to configure and manage the automated processes. This leads to everyone in your business reusing the software, which will lead to faster growth, better utilization and quicker benefits.

2. Framework:
   The kind of architecture the product possesses will determine the way your company uses it. It determines where and how you can use it, and the skills you might need to manage it. The kind of framework you choose depends on the type of work you need the RPA to accomplish. The framework determines the product's features and uses, which directly impacts the growth of your business.

3. Integration:
   One of the main capabilities of RPA is that it can

integrate with a company's current infrastructure easily. However, not every product can smoothly integrate with *your* ERP systems. Each product has different levels of integration. Smooth integration is essential for a quicker, more robust and successful automation. So, when choosing an RPA vendor, verify if their product integrates with your infrastructure seamlessly.

4. Anomaly Detection & Handling:
   When there is an enormous influx of data, there is highly likely to be a few anomalies or errors. In such a scenario, RPAs can detect an anomaly, evade it, and when that is not possible, even solve the anomaly without any human intervention. When the program is designed to necessitate any human action, those particular instances are sent to a separate queue and is visible in reports. Such deft and lucid handling of actions is vital for a robust process to take place and ensures the work goes on in a smooth and steady manner.

5. Configuration:
   Every RPA software have special features embedded in it to help simplify and hasten the process to tweak and edit the configurations of the software. This is done to customize the software to your business's requirements. Good

configuration features ensure impactful use of the automation and support to your infrastructure. Every RPA product has its own set of configuration tools, all of which are distinct in their uses and purposes. Choosing the product that has the features your business requires is essential.

6. Security:
   The deployment of RPA most probably means that there is some degree of sensitive data involved. This is especially important to prevent data leaks that can seriously hamper the image of your company and can even lead to the loss of your business. After all, if you can' safeguard your own data, you probably can't safeguard your customers' data. Currently, the products in the market address various industries that may need various levels of security. That is why finding the RPA product that provides you of the level of protection your business requires is an absolute must.

7. Features:
   After the stage of configuration and testing comes the stage of deployment. The deployment stage entails releasing the product across many systems. This requires the product to allow for handling context-specific factors and offering controls that allow nuanced guidance during

release. The RPA product must have powerful features as various businesses endure various deployment environments.

8. Support and Documentation:
As every business requirements vary, the RPA vendor you choose should provide with excellent customer service, support and documentation. You should be looking for a vendor that provides a strong and robust customer support as well as detailed documentation.

This can help reducing delays, maintenance time and help you provide the best service to your customers. As different RPA vendors offer different kinds of support, do some research before zeroing in on one that you like. However, for the best decision, it's best to hire an IT consultant with RPA experience who will be able to walk you through and able the business to acquire the best fit for the automation process.

This checklist points out some of the major guidelines to look out for when finding an RPA vendor for their business. Another aspect to keep in mind is the budget. Many businesses can't afford an RPA tool that may suit their needs. In other situations, the budget may be a good fit but you might be making a compromise with the chosen tool. Some businesses may have other factors to consider. When making such an important decision, IT consultants with RPA experience and expertise will be the best advisor to guide and help you in this matter. That way, you can get a fully tailor-made solution that will be perfect for your company and its needs.

When zooming in on RPA tools, you need to look at the automation profile and requisites of your business and find a tool that closely matches that. While almost all the RPA vendors claim to be a "one size fits all" kind of tool, there is no RPA solution or tool that can truly be like that. It is really more of a "if the shoe fits" kind of scenario. The key to this conundrum is to find the RPA vendor and tool that closely matches your business requirements. After that, the next step is to customize the RPA tool to the point where it truly "fits like a glove".

The Future of RPA

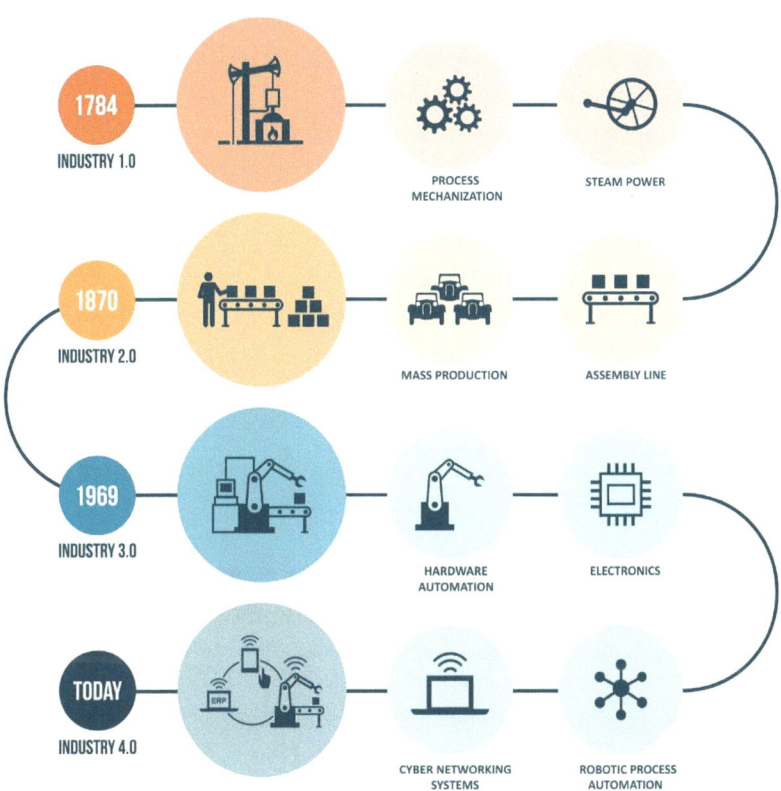

Digital economy is basically an economy that is based on digital computing technologies. As the world is increasing moving towards digital technologies, there is a paramount recognition of the digital economy.

In the digital economy, there has been a disruptive trend going on with the way work gets done – and it is called RPA. In the past few decades, companies had outsourced high volume back office tasks and IT operations, taking advantage of cheap foreign labor. Today, the same can't be said. The work was usually offshored to developing countries like India and Philippines. Today, labor costs have sharply increased. So, the profits of outsourcing work have drastically reduced.

The strong contender to the alternative to off-shored labor is software-run digital labor. This is much cheaper, faster and more efficient than outsourced labor ever was. As it happens, the complete contractual value for outsourced services has declined over the years. Major companies from various industries have already shifted from outsourced human labor to automated software services for high volume, highly transactional manual processes. RPA provides many undeniable benefits like boosting employee productivity. The reason for this is that since software robots increasingly take on the majority of rules-based, repetitive and dangerous tasks, it leaves employees more time to do special value-added activities that can drive customer loyalty and long-term customer value. While many critics and sceptics call RPA a passing trend that won't stay long, RPA is fast becoming a key success factor for companies that want to engage and win in this increasingly competitive digital economy. While the last decade was about obtaining cheaper labor, this coming decade will be about replacing cheaper labor with RPA and smart AI.

References

1. https://en.wikipedia.org/wiki/Industrial_Revolution

2. https://en.wikipedia.org/wiki/Robotic_process_automation

3. https://www2.deloitte.com/content/dam/Deloitte/dk/Documents/finance/FinansAgenda2016/Zeeshan%20Rajan_Danske_Bank_Robotic%20Process%20Automation.pdf

4. https://hbr.org/2017/04/thinking-through-how-automation-will-affect-your-workforce

5. https://en.wikipedia.org/wiki/Fourth_Industrial_Revolution

6. http://www.ey.com/Publication/vwLUAssets/EY-robotic-process-automation-for-hr-and-payroll/$FILE/EY-robotic-process-automation-for-hr-and-payroll.pdf

More Books from The Same Author

How to Start and Run IT Consultancy Business?

Become a Consultant, IT Entrepreneur or Start an Information Technology Consulting Firm

https://www.amazon.com/How-Start-Run-Consultancy-Business-ebook/dp/B01J58E8PK/

Step-By-Step Guide to Start Up a Consulting Business:

Entrepreneurial Small Business Series (Book 1)

https://www.amazon.com/Step-Guide-Start-Consulting-Business-ebook/dp/B01M7NPYLT/

How to Build and Grow Your IT Consulting Startup:

Insider's guide to successful consultancy business startup (Book 2)

https://www.amazon.com/Build-Grow-Your-Consulting-Startup-ebook/dp/B01LZ9ZDOE/

About Author

Srikanth Merianda is a developer, entrepreneur, and investor with over 22 years of experience as an employee and an employer. He has worked for Convergys, Nortel Networks and Blackberry as a Software Engineer, Project Manager, Architect, and Implementation of mission critical projects.

Srikanth has successfully started multiple IT Consulting firms, Business Process Outsource (BPO) firms, Robotic Process Automation, Mobile App Development and Support firms. He has guided these companies through start-up, survival, turnaround and growth phases.

Besides setting up firms, Srikanth also provides IT consulting services to mission critical integration projects as well as technical and cost assessment for software projects. With his wealth of experience, knowledge, and expertise, he has provided technical and business consulting services to companies, some of which have grossed over $1 million in their first year of getting consultancy services from him.

Srikanth is a passionate programmer and an enthusiastic learner of all things technology. He constantly looks for ways to make work and life easier with the tools provided by technology. He holds a Bachelors Engineering degree in

Computer Science and a Master of Science degree in Computer Science from the Mississippi State University.

Author Page

https://www.amazon.com/Srikanth-Merianda/e/B01M5INNL5/

Consulting Opportunity Page

https://www.consultingopportunity.com/

Kiwa. K is a content producer with a lot of curiosity, creativity and experience. With her 4+ years of experience as a content writer, she has ghost-written many fiction and non-fiction bestsellers. She also writes books and delivers engaging content in the technical field. She handles end-to-end content processes for books, eBooks, websites, journals, whitepapers, keynote presentations, articles and blogs.

www.ingramcontent.com/pod-product-compliance
Lightning Source LLC
Chambersburg PA
CBHW040331220526
45473CB00009B/2651